Zack has a big black truck.

He will stack logs in the back.

Zack packs in ten logs.

He stacks four logs on four logs.

He stacks two logs on top.

He gets into his big black truck.

Zack's dog Huck is in the truck.

The two will go to the mill.

Zack's truck hits a big bump.
The big black truck is stuck!
It has a flat.

Zack has a jack.

Click, click, clack, click!

The jack lifts up the truck.

Zack is back in his truck.

He and Huck go to the mill.

Zack drops off ten logs
at the mill.
He drops off two logs,
two logs, two logs,
two logs, and two logs!

Zack and Huck go back.
Zack will have a snack
and a nap.
So will Huck!